Henry M. Johnson

The Iron Manual withTables Showing the Size and Weight of Iron, Steels, Tin Plates, etc.

Henry M. Johnson

The Iron Manual withTables Showing the Size and Weight of Iron, Steels, Tin Plates, etc.

ISBN/EAN: 9783337338916

Printed in Europe, USA, Canada, Australia, Japan

Cover: Foto ©berggeist007 / pixelio.de

More available books at **www.hansebooks.com**

THE

IRON MANUAL;

WITH TABLES,

SHOWING THE SIZE AND WEIGHT OF

IRON, STEEL, TIN PLATES, &c.

Entered, according to Act of Congress, in the year 1868, by

HENRY M. JOHNSON,

In the Clerk's Office of the District Court of the District of Mass.

———◦◇◦———

BOSTON, MASS.
1868.

IRON,

ON account of its abundance, working qualities, and tenacity, is probably the most useful and valuable of metals.

In its primitive position it is commingled with the earth's strata in bountiful profusion. It is found in various combinations and conditions in every formation, and it is a constituent element of both animals and vegetables.

THE ORES.

The ores of iron are found in profuse abundance in every latitude. Imbedded in, or stratified with every formation, they occur both crystallized, massive, and arenaceous; lying deep in strata of vast extent, filling veins and faults in other rocks, and scattered over the surface of the ground. Sometimes, but rarely, found native; usually as oxides, sulphurets, or carbonates, more or less mingled with other substances.

CONVERSION OF CRUDE INTO MALLEABLE IRON.

The conversion of the carbonized crude iron, obtained from the blast furnace, into malleable or

wrought iron, is effected by several operations of an
oxidizing character, in which it is sought to separate,
in the gaseous state, the carbon contained in the
iron by combining it with oxygen, whilst the other
metals, alloyed with the iron and the phosphorus,
pass into the slag.

The iron produced in the smelting furnace may be
divided into two kinds, — that reduced by charcoal,
and that reduced by coke, or raw coal. When charcoal iron has to be converted by charcoal, — as in
Sweden, — it is decarbonized in the charcoal refinery,
with or without an intervening process. Where coal
can be obtained, however, it is now usually converted by the process of puddling. Pig iron, produced by coke or coal, is converted into malleable
iron either by decarbonization in the refinery, or
oxidizing hearth, and subsequent puddling, or it is
converted at once in the puddling furnace by the
process of boiling, — which is equally effective, and
is now more generally practised.

The crude pig iron is assorted according to the de-
gree and uniformity of its carbonization, and classed
as numbers 1, 2, 3, &c.; No. 1 being most highly
carbonized, No. 2 less so, and so on to No. 4, which
contains much more oxygen than the others. The
carbon combined with iron gives it fusibility and
fluidity, but deprives it of ductility. To render it
malleable, and capable of being welded, it must be
deprived, as far as possible, of all the extraneous
substances which have been mixed with it in the
blast furnace, more especially of the carbon.

Prima facie, therefore, it would appear that the
highly carbonized pig iron is the most suitable for

casting, whilst that containing least carbon is best adapted for conversion into malleable iron. Hence, in the trade, the crude iron is divided into foundery and forge pigs.

The pigs, however, in which carbon most predominates, and which, as a rule, have been most effectually separated from all other impurities during the process of smelting, are in many respects preferable for the manufacture of wrought iron. Up to this time, however, great practical difficulties have attended the decarbonization of iron containing so much carbon, and the white or forge iron is almost always preferred, measures having been taken for depriving it of the metals and earthy impurities not separated in the blast furnace.

With regard to the process of refining, we may observe that the crude iron is melted in a hollow fire, and partially decarbonized by the action of a blast of air forced over its surface by a fan or blowing engine; the carbon having greater affinity for the oxygen than for the iron, combines with it, and passes off as gaseous carbonic oxide, or carbonic acid. During this process, a portion of the silicum, &c., is fused out, and separated from the iron. It is obvious from the above, that the iron to be refined, being placed in contact with fuel at a high temperature, is liable to be deteriorated by the admixture of sulphur and other impurities of the fuel; and, as the iron is only partially exposed to the action of the blast, the operation is necessarily, under these circumstances, imperfect. From the refinery the metal is run out into large moulds, and is then broken up into what is technically distinguished as *plate metal*.

The process of puddling succeeds that of refining; and in this operation the reverberatory furnace is employed, with the fire separated by a partition or bridge from the hearth, on which is placed the metal to be puddled. By this arrangement the flame is conducted over the surface of the metal, creating an intense heat ; though the deleterious portions of the fuel cannot mix with the iron in this **furnace, the iron is kept** in a state of fusion, whilst the **workman,** called the "*puddler*," by means of a rake, or rabble, agitates the metal so as to expose, as far as he is able, the whole of the charge to the **action of** the oxygen passing over it from the fire. By this means the carbon is oxidized, and the metal is gradually reduced to a tough, pasty condition, and subsequently to a granular form, somewhat resembling heaps of boiled rice, with the grains greatly enlarged. In this condition of the furnace the cinders, or earthy impurities, yield to the intense heat, and flow **off** from the mass over the bottom in a highly fluid state.

The iron, at this stage, is comparatively pure, and quickly becomes capable of agglutination.

The puddler then collects the metallic granules or particles with his rabble, and rolls them together, backwards and forwards, over the **furnace bottom,** into balls of convenient dimensions (about the size of 13-inch shells), when he removes them from the furnace, to be subjected to the action of the hammer, or mechanical pressure necessary to give the iron homogeneity and fibre.

These processes of refining and puddling have universally been employed till recently; but improvements have rendered it simpler, and the refining process is now very generally abolished.

Shortly after the employment of the puddling process, it was found advantageous to mix a portion of crude iron with the refined plate metal, the expense of the process of refining being saved upon the iron used in the crude state; and, trusting to the decarbonizing effects of the puddling furnace, it was found that the refining process might be altogether dispensed with, if the crude iron containing a portion of oxygen and a very little carbon, was employed. In this single process, it is to be observed, that as all the carbon has to be got rid of in the puddling furnace, the evolution of gas is much more violent, the fluid iron boiling and bubbling energetically during the period of its disengagement; and hence the operation has acquired the popular name of the " boiling " process.

In this operation the pig iron, when melted, is more fluid, — on account of containing a greater proportion of carbon, — than the metal from the refinery, and requires more labor in stirring it about and submitting it to the action of the current of air. The process, moreover, is attended by a greater waste of iron than puddling either plate or crude iron and plate mixed, but not so great a loss as in the two operations of refining and puddling. It must, however, be admitted that the superior fluidity of the iron in the boiling process has a more injurious action on the furnace.

Notwithstanding these objections, the system of boiling, without the intermediate process of refining, has been gaining ground for the last ten years, and in many places has entirely superseded the use of the refinery.

Recent events have therefore led to the conclusion, that in a short time the refining process will have become a thing of the past.

RUSSIA SHEET IRON

Measures 56 by 28 inches, and is rated by the weight per sheet. The numbers run from 8 to 18 Russian pounds per sheet. Eight Russian pounds equal 7.2 English pounds; $9 = 8.1$ lbs.; $10 = 9$ lbs.; $11 = 10$ lbs.; $12 = 11.2$ lbs., &c. 100 Russian pounds equal 90 pounds English.

GALVANIZED IRON,

IRON alloyed superficially with zinc, by plunging the metal, previously well cleaned by friction with dilute acid, into a bath of melted zinc, covered with sal-ammoniac, and stirring it about for some time, produces that which is known as *Galvanized Iron.*

When iron thus treated is exposed to humidity, the zinc is said to become oxidized in consequence of galvanic action. This coating protects the iron beneath from rusting; and hence galvanized iron will retain its whiteness for a long period, under circumstances that would cause ordinary **tinned** iron to exhibit **marks of corrosion.**

To Compute the Weight of Cast Metal by the Weight of the Pattern.

When the Pattern is of White Pine.

RULE. — Multiply the weight of the pattern in pounds by the following multiplier, and the product will give the weight of the casting :

Iron, 14; Brass, 15; Lead, 22; Tin, 14; Zinc, 13.5.

When there are Circular Cores or Prints.

Multiply the square of the diameter of the core or print by its length in inches, the product by .0175, and the result is the weight of the pattern of the core or print, to be deducted from the weight of the pattern.

It is customary, in the making of patterns for castings, to allow for shrinkage per lineal foot of pattern :

Iron and Lead, $\frac{1}{8}$th of an inch, Brass and Zinc, $\frac{8}{16}$ths, and Tin, $\frac{1}{12}$th.

Composition for Welding Cast Steel.

Borax, 10 parts; Sal-ammoniac, 1 part. Grind or pound them roughly together; fuse them in a metal pot over a clear fire, continuing the heat until all spume has disappeared from the surface. When the liquid is **clear**, pour the composition **out to cool** and concrete, and grind to **a** fine powder; then it is ready for use.

To use this composition, the steel to be welded should be raised to a bright yellow heat; then dip it **in** the welding powder, and again raise it to a like **heat as before;** it is then ready to be submitted to **the hammer.**

Shrinkage of Castings.

Iron, small cylinders . . . $=\frac{1}{16}$ inches per **foot.**

"   Pipes . . . . $=\frac{1}{8}$ "   "   "

"   Girders, beams, etc. . $=\frac{1}{8}$ in 15 inches.

"   **Large** cylinders, the contraction of diameter at top . . . $\left.\right\} = \frac{1}{16}$ per foot.

"   Ditto at bottom . . . $=\frac{1}{12}$ per foot.

"   Ditto, in length . . . $=\frac{1}{5}$ in 16 inches.

Brass, thin . . . . $=\frac{1}{8}$ in 9 inches.

"   thick . . . . $=\frac{1}{8}$ in 10 inches.

Zinc . . . . $=\frac{5}{16}$ in a foot.

Lead . . . . $=\frac{5}{16}$ in a foot.

Copper . . . . $=\frac{3}{16}$ in a foot.

Bismuth . . . . $=\frac{5}{32}$ in a **foot.**

Fluxes for Soldering or Welding.

Iron . . . . Borax.
Tinned Iron . . . . Resin.
Copper and Brass . . . . . . Sal-ammoniac.
Zinc . . . . Chloride of zinc.
Lead . . . . Tallow or resin.
Lead and tin pipes . . . . . . Resin and sweet oil.

Steel — Sal-ammoniac, **1** part; borax **10** parts.
Pound together, and fuse until clear, and, when cool,
reduce to powder.

Babbitt's Anti-Attrition Metal.

Melt 4 lbs. Copper; add, by degrees, 12 lbs. best
Banca Tin, 8 lbs. Regulus of Antimony, and 12 lbs.
more of Tin. After 4 or 5 lbs. Tin have been added,
reduce the heat to a dull red, then add the remainder
of the metal as above.

This composition is termed *hardening*; for lining,
take 1 lb. of this *hardening*, melt with it 2 lbs. Banca
Tin, which produces the lining metal for use. Hence,
the proportions for lining metal are 4 lbs. of Copper,
8 of Regulus of Antimony, and 96 of Tin.

To Prevent Iron from Rusting.

Warm it; then rub with white wax; put it again
to the fire until the wax has pervaded the entire
surface.

Or, immerse tools or bright work in boiled linseed-
oil, and allow it to dry upon them.

Weight of Round Rolled Iron, one foot in length.

Size	Weight	Size	Weight	Size	Weight
1/16	.010	2	10.616	5 3/8	76.700
1/8	.041	2 1/8	11.988	5 1/2	80.304
3/16	.094	2 1/4	13.440	5 5/8	84.001
1/4	.165	2 3/8	14.975	5 3/4	87.776
5/16	.261	2 1/2	16.688	5 7/8	91.634
3/8	.373	2 5/8	18.293	6	95.552
7/16	.508	2 3/4	20.076	6 1/4	103.704
1/2	.663	2 7/8	21.944	6 1/2	112.160
9/16	.840	3	23.888	6 3/4	120.960
5/8	1.043	3 1/8	25.926	7	130.048
11/16	1.255	3 1/4	28.040	7 1/4	139.544
3/4	1.493	3 3/8	30.240	7 1/2	149.328
13/16	1.752	3 1/2	32.512	7 3/4	159.456
7/8	2.032	3 5/8	34.886	8	169.856
15/16	2.333	3 3/4	37.332	8 1/4	180.696
1	2.654	3 7/8	39.864	8 1/2	191.808
1 1/16	2.997	4	42.464	8 3/4	203.260
1 1/8	3.360	4 1/8	45.174	9	215.040
1 3/16	3.744	4 1/4	47.952	9 1/4	227.152
1 1/4	4.172	4 3/8	50.815	9 1/2	239.600
1 5/16	4.573	4 1/2	53.760	9 3/4	252.376
1 3/8	5.019	4 5/8	56.788	10	267.008
1 7/16	5.486	4 3/4	59.900	10 1/4	278.924
1 1/2	5.972	4 7/8	63.094	10 1/2	292.688
1 5/8	7.010	5	66.752	11	321.216
1 3/4	8.128	5 1/8	69.731	11 1/2	351.104
1 7/8	9.333	5 1/4	73.172	12	382.208

Weight of Square Rolled Iron, one foot in length.

HENRY M. JOHNSON,
NORWAY AND SWEDES IRON, RODS AND SHAPES,
Boston, Mass.
No. 36 India Street,

1/16	.013	2 3/8	19.066	5 3/4	111.756
1/8	.053	2 1/2	21.120	5 7/8	116.671
3/16	.119	2 5/8	23.292	6	121.604
1/4	.211	2 3/4	25.560	6 1/4	132.040
5/16	.330	2 7/8	27.939	6 1/2	142.816
3/8	.475	3	30.416	6 3/4	154.012
7/16	.647	3 1/8	33.010	7	165.632
1/2	.845	3 1/4	35.704	7 1/4	177.672
9/16	1.069	3 3/8	38.503	7 1/2	190.136
5/8	1.320	3 1/2	41.408	7 3/4	203.024
11/16	1.597	3 5/8	44.418	8	216.336
3/4	1.901	3 3/4	47.534	8 1/4	230.068
13/16	2.231	3 7/8	50.756	8 1/2	244.220
7/8	2.588	4	54.084	8 3/4	258.800
15/16	2.971	4 1/8	57.517	9	273.792
1	3.380	4 1/4	61.055	9 1/4	289.220
1 1/16	3.816	4 3/8	64.700	9 1/2	305.056
1 1/8	4.278	4 1/2	68.448	9 3/4	321.332
1 1/4	5.280	4 5/8	72.305	10	337.920
1 3/8	6.390	4 3/4	76.264	10 1/4	355.136
1 1/2	7.604	4 7/8	80.333	10 1/2	372.672
1 5/8	8.926	5	84.480	10 3/4	390.628
1 3/4	10.352	5 1/8	88.784	11	408.960
1 7/8	11.883	5 1/4	93.168	11 1/4	427.812
2	13.520	5 3/8	97.657	11 1/2	447.024
2 1/8	15.263	5 1/2	102.240	11 3/4	466.684
2 1/4	17.112	5 5/8	106.953	12	486.656

Weight of Flat Rolled Iron, one Foot in Length.

Width.	Thickness.				
	$\frac{1}{4}$	$\frac{5}{16}$	$\frac{3}{8}$	$\frac{7}{16}$	$\frac{1}{2}$
$\frac{1}{2}$	.422	.528	.634	.738	.845
$\frac{5}{8}$	.528	.660	.792	.923	1.056
$\frac{3}{4}$	.633	.792	.950	1.108	1.265
$\frac{7}{8}$	.738	.923	1.108	1.294	1.477
1	.843	1.056	1.267	1.478	1.690
$1\frac{1}{8}$	.950	1.187	1.425	1.663	1.901
$1\frac{1}{4}$	1.056	1.320	1.584	1.848	2.112
$1\frac{3}{8}$	1.161	1.452	1.742	2.032	2.325
$1\frac{1}{2}$	1.266	1.584	1.900	2.217	2.535
$1\frac{5}{8}$	1.372	1.716	2.059	2.402	2.746
$1\frac{3}{4}$	1.479	1.848	2.218	2.589	2.957
$1\frac{7}{8}$	1.584	1.980	2.376	2.772	3.168
2	1.689	2.112	2.534	2.957	3.379
$2\frac{1}{8}$	1.795	2.244	2.693	3.141	3.591
$2\frac{1}{4}$	1.900	2.376	2.851	3.326	3.802
$2\frac{3}{8}$	2.006	2.508	3.009	3.511	4.013
$2\frac{1}{2}$	2.112	2.640	3.168	3.696	4.224
$2\frac{3}{4}$	2.323	2.904	3.485	4.066	4.647
3	2.535	3.168	3.802	4.435	5.069
$3\frac{1}{4}$	2.746	3.432	4.119	4.805	5.492
$3\frac{1}{2}$	2.957	3.696	4.436	5.175	5.914
$3\frac{3}{4}$	3.168	3.960	4.752	5.544	6.336
4	3.380	4.224	5.069	5.914	6.759
$4\frac{1}{2}$	3.802	4.752	5.703	6.653	7.604
5	4.224	5.280	6.336	7.392	8.449
$5\frac{1}{2}$	4.647	5.808	6.970	8.132	9.294
6	5.070	6.337	7.604	8.871	10.138

Weight of Flat Rolled Iron, one Foot in Length.

HENRY M. JOHNSON,
NORWAY AND SWEDES IRON, RODS AND SHAPES,
Boston, Mass.
No. 36 India Street,

Width.	Thickness.				
	5/8	3/4	7/8	1	1¼
½	1.056	1.265	1.477	1.690	2.112
⅝	1.320	1.584	1.846	2.112	2.640
¾	1.584	1.901	2.217	2.534	3.168
⅞	1.846	2.217	2.588	2.956	3.696
1	2.112	2.534	2.956	3.380	4.224
1⅛	2.375	2.850	3.326	3.802	4.752
1¼	2.640	3.168	3.696	4.224	5.280
1⅜	2.904	3.484	4.065	4.646	5.808
1½	3.168	3.802	4.435	5.069	6.337
1⅝	3.432	4.119	4.805	5.492	6.864
1¾	3.696	4.435	5.178	5.914	7.393
1⅞	3.960	4.752	5.544	6.336	7.921
2	4.224	5.069	5.914	6.758	8.448
2⅛	4.488	5.386	6.283	7.181	8.977
2¼	4.752	5.703	6.653	7.604	9.505
2⅜	5.016	6.019	7.022	8.025	10.032
2½	5.280	6.336	7.392	8.448	10.560
2¾	5.808	6.970	8.132	9.294	11.617
3	6.337	7.604	8.871	10.138	12.673
3¼	6.865	8.237	9.610	10.983	13.730
3½	7.393	8.871	10.350	11.828	14.785
3¾	7.921	9.505	11.089	12.673	15.841
4	8.448	10.138	11.828	13.518	16.897
4½	9.504	11.406	13.306	15.208	19.010
5	10.560	12.673	14.784	16.897	21.122
5½	11.616	13.940	16.264	18.587	23.234
6	12.674	15.208	17.742	20.276	25.346

Weight of Wrought Angle Iron, one Foot in Length,

Thickness Measured in the Middle of each Side.

EQUAL SIDES.

Sides.	Thickness.	Weight.
Inches.	Inches.	Pounds.
1.25 × 1.25	$\frac{3}{16}$	1.5
1.5 × 1.5	$\frac{3}{16}$	2.
1.75 × 1.75	$\frac{1}{4}$	3.
2. × 2.	$\frac{1}{4}$	3.5
2.25 × 2.25	$\frac{5}{16}$	4.5
2.5 × 2.5	$\frac{5}{16}$	5.
3. × 3.	$\frac{3}{8}$	7.
3.5 × 3.5	$\frac{7}{16}$	9.
4. × 4.	$\frac{1}{2}$	12.5
4.5 × 4.5	$\frac{1}{2}$	14.
4.5 × 4.5	$\frac{9}{16}$	16.

UNEQUAL SIDES.

Sides.	Thickness.	Weight.
3.5 × 3.	$\frac{7}{16}$	9.6
4. × 3.	$\frac{1}{2}$	11.
4. × 3.5	$\frac{1}{2}$	11.5
4. × 3.5	$\frac{9}{16}$	11.75
4.5 × 3.	$\frac{9}{16}$	11.75
5. × 3.	$\frac{1}{2}$	12.65
5. × 3.	$\frac{9}{16}$	13.7
5.5 × 3.5	$\frac{1}{2}$	14.5
5.5 × 3.5	$\frac{9}{16}$	15.6
6. × 3.5	$\frac{5}{8}$	18.
6. × 4.5	$\frac{5}{8}$	20.

Whale or Oil Cask Hoops.

Weight.	Penny.	Width.	W. G.	W't. pr Ft.
Light	3d.	1⅛	14	.3122
Heavy	3d.	1⅛	13	.3574
Light	4d.	1¼	13	.3971
Heavy	4d.	1⅜	12	.5011
Light	5d.	1½	11	.6019
Heavy	5d.	1⅝	10	.7281
Light	6d.	1¾	10	.7841
Heavy	6d.	1¾	9	.8660

Table of the Thickness and Weight of Galvanized Sheet Iron.

Dimensions of Sheet, 2 to 3 Feet in Width by from 6 to 9 Feet in Length.

Wire Gauge.	W'ght per Sq. Foot.	Wire Gauge.	W'ght per Sq. Foot.
Number.	Ounces.	Number.	Ounces.
30	10	22	21
29	11	21	24
28	12	20	28
27	14	19	33
26	15	18	37
25	16	17	43
24	17	16	48
23	19	14	60

Weight of Hoop and Band Iron, one Foot in
Length — Birmingham Wire Gauge.

Width.	W.G. 22	W.G. 21	W.G. 20	W.G. 19	W.G. 18
$\frac{1}{2}$	.0468	.0535	.0585	.0702	.0819
$\frac{5}{8}$	.0585	.0668	.0731	.0877	.1024
$\frac{3}{4}$	.0702	.0802	.0878	.1053	.1229
$\frac{7}{8}$	.0819	.0936	.1024	.1228	.1434
1	.0936	.1070	.1170	.1404	.1638
$1\frac{1}{8}$	.1053	.1204	.1316	.1579	.1843
$1\frac{1}{4}$	.1170	.1337	.1463	.1755	.2048
$1\frac{3}{8}$	.1287	.1471	.1609	.1930	.2253
$1\frac{1}{2}$	.1404	.1605	.1755	.2106	.2458
$1\frac{5}{8}$	.1521	.1739	.1901	.2281	.2663
$1\frac{3}{4}$	.1638	.1872	.2048	.2457	.2867
$1\frac{7}{8}$	.1755	.2006	.2194	.2632	.3072
2	.1872	.2140	.2341	.2808	.3276
$2\frac{1}{4}$	.2106	.2407	.2633	.3159	.3686
$2\frac{1}{2}$	.2340	.2675	.2926	.3510	.4096
$2\frac{3}{4}$	.2574	.2942	.3219	.3861	.4506
3	.2808	.3210	.3510	.4212	.4914

Weight of Hoop and Band Iron, one Foot in
Length — Birmingham Wire Gauge.

	Width.	W.G. 17	W.G. 16	W.G. 15	W.G. 14	W.G. 13
HENRY M. JOHNSON, NORWAY AND SWEDES IRON, RODS AND SHAPES, Boston, Mass. No. 36 India Street,	$\frac{1}{2}$	.0970	.1087	.1204	.1388	.1588
	$\frac{5}{8}$	.1212	.1359	.1505	.1735	.1985
	$\frac{3}{4}$	.1455	.1630	.1806	.2081	.2382
	$\frac{7}{8}$	.1697	.1902	.2107	.2428	.2779
	1	.1939	.2173	.2407	.2775	.3177
	$1\frac{1}{8}$	.2182	.2445	.2708	.3122	.3574
	$1\frac{1}{4}$	.2424	.2717	.3009	.3469	.3971
	$1\frac{3}{8}$	.2667	.2988	.3310	.3816	.4368
	$1\frac{1}{2}$	.2909	.3260	.3611	.4163	.4765
	$1\frac{5}{8}$	.3151	.3532	.3912	.4510	.5162
	$1\frac{3}{4}$	.3394	.3803	.4213	.4857	.5559
	$1\frac{7}{8}$	.3636	.4075	.4514	.5204	.5956
	2	.3879	.4347	.4815	.5551	.6353
	$2\frac{1}{4}$	.4364	.4890	.5417	.6245	.7147
	$2\frac{1}{2}$	.4849	.5434	.6019	.6939	.7941
	$2\frac{3}{4}$	.5334	.5977	.6621	.7632	.8735
	3	.5818	.6520	.7222	.8326	.9530

Weight of Hoop and Band Iron, one Foot in Length — Birmingham Wire Gauge.

	W.G. 12	W.G. 11	W.G. 10	W.G. 9	W.G. 8
Width.					
$\frac{1}{2}$	.1822	.2006	.2240	.2474	.2759
$\frac{5}{8}$	.2278	.2507	.2800	.3093	.3448
$\frac{3}{4}$	.2733	.3009	.3360	.3711	.4138
$\frac{7}{8}$	.3189	.3511	.3920	.4330	.4828
1	.3645	.4012	.4480	.4940	.5517
$1\frac{1}{8}$	.4100	.4514	.5040	.5568	.6207
$1\frac{1}{4}$	.4556	.5015	.5600	.6186	.6896
$1\frac{3}{8}$	.5011	.5517	.6160	.6805	.7586
$1\frac{1}{2}$	.5467	.6019	.6721	.7423	.8276
$1\frac{5}{8}$	.5922	.6520	.7281	.8042	.8966
$1\frac{3}{4}$	.6378	.7022	.7841	.8660	.9655
$1\frac{7}{8}$	.6834	.7523	.8401	.9279	1.0341
2	.7289	.8025	.8961	.9897	1.1034
$2\frac{1}{4}$	.8200	.9028	1.0081	1.1134	1.2413
$2\frac{1}{2}$	.9111	1.0031	1.1201	1.2371	1.3793
$2\frac{3}{4}$	1.0022	1.1034	1.2321	1.3608	1.5172
3	1.0934	1.2037	1.3442	1.4846	1.6551

Weight of Hoop and Band Iron, one Foot in Length — Birmingham Wire Gauge.

HENRY M. JOHNSON, NORWAY AND SWEDES IRON, RODS AND SHAPES, Boston, Mass. No. 36 India Street,

Width.	W.G. 7	W.G. 6	W.G. 5	W.G. 4	W.G. 3
$\frac{1}{2}$	.3010	.3394	.3678	.3978	.4330
$\frac{5}{8}$	.3762	.4242	.4597	.4973	.5412
$\frac{3}{4}$	.4514	.5091	.5517	.5968	.6495
$\frac{7}{8}$	.5267	.5939	.6436	.6963	.7577
1	.6020	.6788	.7356	.7958	.8660
$1\frac{1}{8}$	.6772	.7636	.8275	.8952	.9743
$1\frac{1}{4}$	.7525	.8485	.9195	.9947	1.0825
$1\frac{3}{8}$	.8277	.9333	1.0114	1.0942	1.1908
$1\frac{1}{2}$	.9028	1.0182	1.1034	1.1937	1.2990
$1\frac{5}{8}$	.9780	1.1030	1.1953	1.2932	1.4072
$1\frac{3}{4}$	1.0533	1.1879	1.2873	1.3926	1.5155
$1\frac{7}{8}$	1.1285	1.2727	1.3792	1.4922	1.6237
2	1.2037	1.3576	1.4712	1.5916	1.7321
$2\frac{1}{4}$	1.3542	1.5274	1.6551	1.7905	1.9486
$2\frac{1}{2}$	1.5047	1.6070	1.8390	1.9894	2.1651
$2\frac{3}{4}$	1.6551	1.8667	2.0229	2.1884	2.3816
3	1.8056	2.0363	2.2069	2.3874	2.5981

Weights of Wrought Iron and Steel.

Thickness determined by Birmingham Gauge.

No. of Gauge.	Thickness of each No.	Plates — per Sq. Foot.	
		Iron.	Steel.
	Inches.	Pounds.	Pounds.
0000	.454	18.2167	18.4596
000	.425	17.0531	17.2805
00	.38	15.2475	15.4508
0	.34	13.6425	13.8244
1	.3	12.0375	12.198
2	.284	11.3955	11.5474
3	.259	10.3924	10.5309
4	.238	9.5497	9.6771
5	.22	8.8275	8.9452
6	.203	8.1454	8.254
7	.18	7.2225	7.3188
8	.165	6.6206	6.7089
9	.148	5.9385	6.0177
10	.134	5.3767	5.4484
11	.12	4.815	4.8792
12	.109	4.3736	4.4319
13	.095	3.8119	3.8627
14	.083	3.3304	3.3748
15	.072	2.889	2.9275
16	.065	2.6081	2.6429
17	.058	2.3272	2.3583
18	.049	1.9661	1.9923

Weights of Wrought Iron and Steel.

Thickness determined by Birmingham Gauge.

No. of Gauge.	Thickness of each No.	Plates — per Sq. Foot.	
		Iron.	Steel.
	Inches.	Pounds.	Pounds.
19	.042	1.6852	1.7077
20	.035	1.4044	1.4231
21	.032	1.284	1.3011
22	.028	1.1235	1.1385
23	.025	1.0031	1.0165
24	.022	.8827	.8945
25	.02	.8025	.8132
26	.018	.7222	.7319
27	.016	.642	.6506
28	.014	.5617	.5692
29	.013	.5216	.5286
30	.012	.4815	.4879
31	.01	.4012	.4066
32	.009	.3611	.3659
33	.008	.321	.3253
34	.007	.2809	.2846
35	.005	.2006	.2033
36	.004	.1605	.1626

HENRY M. JOHNSON,

NORWAY AND SWEDES IRON, RODS AND SHAPES.

Boston, Mass.

No. 36 India Street,

Weights of Copper and Brass Plates.

Thickness determined by Birmingham Gauge.

No. of Gauge.	Thickness of each No.	Plates — per Sq. Foot.	
		Copper.	Brass.
	Inches.	Pounds.	Pounds.
0000	.454	20.5662	19.4312
000	.425	19.2525	18.19
00	.38	17.214	16.264
0	.34	15.402	14.552
1	.3	13.59	12.84
2	.284	12.8652	12.1552
3	.259	11.7327	11.0852
4	.238	10.7814	10.1864
5	.22	9.966	9.416
6	.203	9.1959	8.6884
7	.18	8.154	7.704
8	.165	7.4745	7.062
9	.148	6.7044	6.3344
10	.134	6.0702	5.7352
11	.12	5.436	5.136
12	.109	4.9377	4.6652
13	.095	4.3035	4.066
14	.083	3.7599	3.5524
15	.072	3.2616	3.0816
16	.065	2.9445	2.782
17	.058	2.6274	2.4824
18	.049	2.2197	2.0972

Weights of Copper and Brass Plates.

Thickness determined by Birmingham Gauge.

No. of Gauge.	Thickness of each No.	Plates — per Sq. Foot.	
		Copper.	Brass.
	Inches.	Pounds.	Pounds.
19	.042	1.9026	1.7976
20	.035	1.5855	1.498
21	.032	1.4496	1.3696
22	.028	1.2684	1.1984
23	.025	1.1325	1.07
24	.022	.9966	.9413
25	.02	.906	.856
26	.018	.8154	.7704
27	.016	.7248	.6848
28	.014	.6342	.5992
29	.013	.5889	.5564
30	.012	.5436	.5136
31	.01	.453	.428
32	.009	.4077	.3852
33	.008	.3624	.3424
34	.007	.3171	.2996
35	.005	.2265	.214
36	.004	.1812	.1712

TIN PLATES.

Mark.	Number of Sheets per Box.	Length and Breadth in Inches.	W'ght per Box.
IC	225	10 × 14	112
IX	225	10 × 14	140
IXX	225	10 × 14	161
IXXX . . .	225	10 × 14	182
IXXXX . .	225	10 × 14	203
IC	112	14 × 20	112
IX	112	14 × 20	140
IXX	112	14 × 20	168
IXXX . . .	112	14 × 20	196
IXXXX . .	112	14 × 20	224
DC	100	12½ × 17	98
DX	100	12½ × 17	126
DXX . . .	100	12½ × 17	147
DXXX . . .	100	12½ × 17	168
DXXXX . .	100	12½ × 17	189
SDC	200	11 × 15	167
SDX	200	11 × 15	188
SDXX . . .	200	11 × 15	209
SDXXX . .	200	11 × 15	230
SDXXXX . .	200	11 × 15	251
IIC	225	9¾ × 13¼	105
IIX	225	9¾ × 13¼	133
IIIC	225	9½ × 12¾	98

Mark.	Number of Sheets per Box.	Leogth and Breadth in Inches.	W'ght per Box.
IIIX . . .	225	9½ × 12¾	126
TT	450	10 × 14	112
XTT . . .	450	10 × 14	126
IC	225	11 × 11	96
IC	225	13 × 13	140
IC	225	14 × 14	168
IXX . . .		12 × 24	
IXXX . .	These sizes are sold at — per lb.	13 × 21	
IXXXX .		14 × 22	
IXXXXX		14 × 26	
IXXXXXX		14 × 26½	
IC	225	12 × 12	115
IX	225	12 × 12	144
IXX . . .	225	12 × 12	166
IXXX . .	225	12 × 12	187
IXXXX .	225	12 × 12	209

TERNE PLATES.

IC	112	14 × 20	112
IX	112	14 × 20	110

Weight of Cast Iron Pipes of different Thicknesses, From 1 Inch to 36 Inches in Diameter.

One Foot in Length.

Diam.	Thick.	W'ght.	Diam.	Thick.	W'ght.
Inches.	Inches.	Pounds.	Inches.	Inches.	Pounds.
1.	¼	3.06	3.¾	½	20.9
1.	⅜	5.05	3.¾	⅝	26.83
1.¼	¼	3.67	3.¾	¾	33.07
1.¼	⅜	6.	4.	½	22.05
1.½	⅜	6.89	4.	⅝	28.28
1.½	½	9.8	4.	¾	34.94
1.¾	⅜	7.8	4.¼	½	23.35
1.¾	½	11.04	4.¼	⅝	29.85
2.	⅜	8.74	4.¼	¾	36.73
2.	½	12.23	4.½	½	24.49
2.¼	⅜	9.65	4.½	⅝	31.4
2.¼	½	13.48	4.½	¾	38.58
2.½	⅜	10.57	4.¾	½	25.7
2.½	½	14.66	4.¾	⅝	32.91
2.½	⅝	19.05	4.¾	¾	40.43
2.¾	⅜	11.54	5.	½	26.94
2.¾	½	15.91	5.	⅝	34.34
2.¾	⅝	20.59	5.	¾	42.28
3.	⅜	12.28	5.½	½	29.4
3.	½	17.15	5.½	⅝	37.44
3.	⅝	22.15	5.½	¾	45.94
3.	¾	27.56	6.	½	31.82
3.¼	½	18.4	6.	⅝	40.56
3.¼	⅝	23.72	6.	¾	49.6
3.¼	¾	20.64	6.	⅞	58.96
3.½	½	19.66	6.½	½	34.32
3.½	⅝	25.27	6.½	⅝	43.68
3.½	¾	31.2	6.½	¾	53.3

TABLE—Continued.

HENRY M. JOHNSON, NORWAY AND SWEDES IRON, RODS AND SHAPES, Boston, Mass. No. 36 India Street,

Diam.	Thick.	W'ght.	Diam.	Thick.	W'ght.
Inches.	Inches.	Pounds.	Inches.	Inches.	Pounds.
6.½	.⅞	63.18	9.½	1.	102.9
7.	.½	36.66	10.	.½	51.46
7.	.⅝	46.8	10.	.⅝	65.08
7.	.¾	56.96	10.	.¾	78.99
7.	.⅞	67.6	10.	.⅞	93.24
7.	1.	78.39	10.	1.	108.84
7.½	.½	39.22	10.½	.½	53.88
7.½	.⅝	49.92	10.½	.⅝	68.14
7.½	.¾	60.48	10.½	.¾	82.68
7.½	.⅞	71.76	10.½	.⅞	97.44
7.½	1.	83.28	10.½	1.	112.68
8.	.½	41.64	11.	.½	56.34
8.	.⅝	52.68	11.	.⅝	71.19
8.	.¾	64.27	11.	.¾	86.4
8.	.⅞	76.12	11.	.⅞	101.83
8.	1.	88.2	11.	1.	117.6
8.½	.½	44.11	11.½	.½	58.82
8.½	.⅝	56.16	11.½	.⅝	74.28
8.½	.¾	68.	11.½	.¾	90.06
8.½	.⅞	80.5	11.½	.⅞	106.14
8.½	1.	93.28	11.½	1.	122.62
9.	.½	46.5	12.	.½	61.26
9.	.⅝	58.92	12.	.⅝	77.36
9.	.¾	71.7	12.	.¾	93.7
9.	.⅞	84.7	12.	.⅞	110.48
9.	1.	97.98	12.	1.	127.42
9.½	.½	48.98	12.½	.½	63.7
9.½	.⅝	62.02	12.½	.⅝	80.4
9.½	.¾	75.32	12.½	.¾	97.4
9.½	.⅞	88.98	12.½	.⅞	114.72

TABLE—Continued.

Diam.	Thick.	W'ght	Diam.	Thick.	W'ght
Inches.	Inches.	Pounds.	Inches.	Inches.	Pounds.
12.½	1.	132.35	15.½	1.	161.82
13.	.½	66.14	16.	.½	80.87
13.	.⅝	83.46	16.	.⅝	101.82
13.	.¾	101.08	16.	.¾	123.14
13.	.⅞	118.97	16.	.⅞	144.76
13.	1.	137.28	16.	1.	166.6
13.½	.½	68.36	17.	.½	85.73
13.½	.⅝	86.55	17.	.⅝	107.96
13.½	.¾	104.76	17.	.¾	130.48
13.½	.⅞	123.3	17.	.⅞	153.3
13.½	1.	142.16	17.	1.	176.58
14.	.½	71.07	18.	.⅝	114.1
14.	.⅝	89.61	18.	.¾	137.84
14.	.¾	108.46	18.	.⅞	161.9
14.	.⅞	127.6	18.	1.	186.24
14.	1.	147.03	19.	.⅝	120.24
14.½	.½	73.72	19.	.¾	145.2
14.½	.⅝	92.66	19.	.⅞	170.47
14.½	.¾	112.1	19.	1.	195.92
14.½	.⅞	131.86	20.	.⅝	126.33
14.½	1.	151.92	20.	.¾	152.53
15.	.½	75.96	20.	.⅞	179.02
15.	.⅝	95.72	20.	1.	205.8
15.	.¾	115.78	21.	.⅝	132.5
15.	.⅞	136.15	21.	.¾	159.84
15.	1.	156.82	21.	.⅞	187.6
15.½	.½	78.4	21.	1.	215.52
15.½	.⅝	98.78	22.	.⅝	138.6
15.½	.¾	119.48	22.	.¾	167.24
15.½	.⅞	140.4	22.	.⅞	196.46

TABLE—Continued.

HENRY M. JOHNSON, NORWAY AND SWEDES IRON, RODS AND SHAPES, No. 36 India Street, Boston, Mass.

Diam.	Thick.	W'ght	Diam.	Thick.	W'ght
Inches.	Inches.	Pounds.	Inches.	Inches.	Pounds.
22.	1.	225.38	31.	.¾	233.4
23.	.⅝	144.77	31.	.⅞	273.4
23.	.¾	174.62	31.	1.	313.68
23.	.⅞	204.78	31.	1.⅛	354.24
23.	1.	235.28	32.	.¾	240.76
24.	.⅝	150.85	32.	.⅞	281.94
24.	.¾	181.92	32.	1.	323.49
24.	.⅞	213.28	32.	1.⅛	365.29
24.	1.	245.08	33.	.¾	248.1
25.	.⅝	156.97	33.	.⅞	290.5
25.	.¾	189.28	33.	1.	333.24
25.	.⅞	221.94	33.	1.⅛	376.26
25.	1.	254.86	33.	1.¼	420.77
26.	.¾	196.62	34.	.¾	255.45
26.	.⅞	230.56	34.	.⅞	298.88
26.	1.	264.66	34.	1.	342.88
27.	.¾	204.04	34.	1.⅛	387.13
27.	.⅞	239.08	34.	1.¼	431.76
27.	1.	274.56	35.	.¾	262.7
28.	.¾	211.32	35.	.⅞	307.62
28.	.⅞	247.62	35.	1.	352.86
28.	1.	284.28	35.	1.⅛	398.1
29.	.¾	218.7	35.	1.¼	443.96
29.	.⅞	256.2	36.	.¾	270.18
29.	1.	294.02	36.	.⅞	316.36
30.	.¾	226.2	36.	1.	362.86
30.	.⅞	264.79	36.	1.⅛	409.34
30.	1.	303.86	36.	1.¼	456.46
30.	1.⅛	343.2			

NOTE.—These weights do not include any allowance for spigot and faucet ends.

Table of Standard Dimensions of Wrought Iron Welded Tubes.

Nominal Diam.	External Diam	Thickness.	Internal Diam.	Internal Circumf.	External Circumf.
Inches.	Inches.	Inches.	Inches.	Inches.	Inches.
⅛	.40	.068	.27	.85	1.27
¼	.54	.088	.36	1.14	1.7
⅜	.67	.091	.49	1.55	2.12
½	.84	.109	.62	1.96	2.65
¾	1.05	.113	.82	2.59	3.3
1	1.31	.134	1.05	3.29	4.13
1¼	1.66	.14	1.38	4.33	5.21
1½	1.9	.145	1.61	5.06	5.97
2	2.37	.154	2.07	6.49	7.46
2½	2.87	.204	2.47	7.75	9.03
3	3.5	.217	3.07	9.64	11.
3½	4.	.226	3.55	11.15	12.57
4	4.5	.237	4.07	12.69	14.14
4½	5.	.247	4.51	14.15	15.71
5	5.56	.259	5.04	15.85	17.47
6	6.62	.28	6.06	19.05	20.81
7	7.62	.301	7.02	22.06	23.95
8	8.62	.322	7.98	25.06	27.1
9	9.69	.344	9.	28.28	30.43
10	10.75	.366	10.02	31.47	33.77

Table of Standard Dimensions of Wrought Iron Welded Tubes.

Nominal Diam.	L'gth of Pipe per Sq. Ft. of Internal Surface.	L'gth of Pipe per Sq. Ft. of External Surface.	Internal Area.	Weight per Foot.	No. of Threads per In. of Screw.
Inches.	Feet..	Feet.	Inches.	Pounds.	
⅛	14.15	9.44	.057	.24	27
¼	10.5	7.075	.104	.42	18
⅜	7.07	5.657	.192	.56	18
½	6.13	4.502	.305	.84	14
¾	4.64	3.637	.533	1.13	14
1	3.66	2.903	.863	1.67	11¼
1¼	2.77	2.301	1.496	2.26	11½
1½	2.37	2.01	2.038	2.69	11½
2	1.85	1.611	3.355	3.67	11½
2½	1.55	1.328	4.783	5.77	8
3	1.24	1.091	7.388	7.55	8
3½	1.08	0.955	9.887	9.05	8
4	.95	0.849	12.73	10.73	8
4½	.85	0.765	15.939	12.49	8
5	.78	0.629	19.99	14.56	8
6	.63	0.577	28.889	18.77	8
7	.54	0.505	38.737	23.41	8
8	.48	0.414	50.039	28.35	8
9	.42	0.394	63.633	34.08	8
10	.38	0.355	78.838	40.64	8

Weight of Composition Sheathing Nails.

No.	L'gth.	No. in a Pound.	No.	L'gth.	No. in a Pound.
	Inches.			Inches.	
1	¾	290	8	1¼	168
2	⅞	260	9	1½	110
3	1	212	10	1⅝	101
4	1⅛	201	11	1¾	74
5	1¼	199	12	2	64
6	1	190	13	2¼	59
7	1⅛	184			

Length of Horseshoe Nails.

No. 5	1½ Ins.		No. 8	2	Ins.
" 6	1¾ "		" 9	2¼	"
" 7	1⅞ "		" 10	2½	"

L'gths of Iron Nails, and No. in Pound.

Size.	Length.	No.	Size.	Length.	No.
3d.	1¼	427	10d.	3	61
4	1½	261	12	3¼	50
5	1¾	200	20	3½	37
6	2	146	30	4	24
8	2½	95	40	4¼	17

Value of Iron per Ton of 2240 Pounds,

At from 2 Cents to 12 Cents per Pound.

HENRY M. JOHNSON, NORWAY AND SWEDES IRON, RODS AND SHAPES, Boston, Mass. No. 36 India Street,

2	44.80	5⅜	120.40	8¾	196.00
2⅛	47.60	5½	123.20	8⅞	198.80
2¼	50.40	5⅝	126.00	9	201.60
2⅜	53.20	5¾	128.80	9⅛	204.40
2½	56.00	5⅞	131.60	9¼	207.20
2⅝	58.80	6	134.40	9⅜	210.00
2¾	61.60	6⅛	137.20	9½	212.80
2⅞	64.40	6¼	140.00	9⅝	215.60
3	67.20	6⅜	142.80	9¾	218.40
3⅛	70.00	6½	145.60	9⅞	221.20
3¼	72.80	6⅝	148.40	10	224.00
3⅜	75.60	6¾	151.20	10⅛	226.80
3½	78.40	6⅞	154.00	10¼	229.60
3⅝	81.20	7	156.80	10⅜	232.40
3¾	84.00	7⅛	159.60	10½	235.20
3⅞	86.80	7¼	162.40	10⅝	238.00
4	89.60	7⅜	165.20	10¾	240.80
4⅛	92.40	7½	168.00	10⅞	243.60
4¼	95.20	7⅝	170.80	11	246.40
4⅜	98.00	7¾	173.60	11⅛	249.20
4½	100.80	7⅞	176.40	11¼	252.00
4⅝	103.60	8	179.20	11⅜	254.80
4¾	106.40	8⅛	182.00	11½	257.60
4⅞	109.20	8¼	184.80	11⅝	260.40
5	112.00	8⅜	187.60	11¾	263.20
5⅛	114.80	8½	190.40	11⅞	266.00
5¼	117.60	8⅝	193.20	12	268.80

INDEX

IRON.

www.ingramcontent.com/pod-product-compliance
Lightning Source LLC
Chambersburg PA
CBHW022033190326
41519CB00010B/1693